BRITISH MILITARY INTELLIGENCE

Objects from the Military Intelligence Museum

Nick van der Bijl

AMBERLEY

First published 2017

Amberley Publishing
The Hill, Stroud
Gloucestershire, GL5 4EP

www.amberley-books.com

Copyright © Nick van der Bijl, 2017

The right of Nick van der Bijl to be identified
as the Author of this work has been asserted in
accordance with the Copyrights, Designs and
Patents Act 1988.

ISBN 978 1 4456 6238 1 (print)
ISBN 978 1 4456 6239 8 (ebook)

British Library Cataloguing in Publication Data.
A catalogue record for this book is available from
the British Library.

Typesetting by Amberley Publishing.
Printed in the UK.

Contents

Introduction & Acknowledgements

This book focuses on telling a history of British military intelligence from the nineteenth century to the most recent campaigns by describing some objects lodged in the Military Intelligence Museum and the Archives at Chicksands in Bedfordshire, and then telling the story behind them or explaining why the object is of historical interest. Some readers may be disappointed at the lack of firearms and military hardware, but this is inevitable in a military skill that relies on gathering information in peace and war and converting it into a product of value to a commander.

I must thank several people. To the Military Intelligence Museum trustees, past and present, for their encouragement. To Major Bill Steadman, the museum curator, and Joyce Hutton, the archivist, and to their staff for their detailed assistance and guidance in a subject that has a considerable number of barriers. To Mrs Evelyn Le Chene for information on her husband, Pierre. To Commander Stephen Foster for supplying information on Lance-Corporal Coulthard. Photographs have been assembled courtesy of the Archives, except where annotated. Most important, I must thank Alan Atkinson for supporting the project with photography and his commitment to the project. To Alexander Stilwell of Amberley Publishing for his advice.

Yet again to my wife, Penny, for her patience and support.

Nick van der Bijl
January 2017

Glossary

ATS	Auxiliary Territorial Service (Women's Branch of the Army 1938–1949)
BAOR	British Army of the Rhine
BEM	British Empire Medal
BEF	British Expeditionary Force
BRIXMIS	British Commanders'-in-Chief Mission to the Soviet Forces in Germany
CI	Counter-Intelligence
CMG	Companion of St Michael and St George
CSM	Company Sergeant Major
DCM	Distinguished Conduct Medal
DL	Deputy Lieutenant
DLB	Dead Letter Box
DSO	Distinguished Service Order
FID	Field Intelligence Department
FSP	Field Security Police
FS	Field Security
FSS	Field Security Section
GFP	*Geheime Feldpolizie* (Field Secret Police)
GCMG	Grand Cross of the Order of St Michael and St George
GCVO	Grand Cross of the (Royal) Victorian Order
GHQ	General Headquarters
GM	George Medal
GRU	*Glavnoye razvedyvatel'noye upravleniye* (Main Intelligence Agency of the General Staff of the Armed Forces of the Soviet Union/Russian Federation)
GSM	General Service Medal
GSO	General Staff Officer
HQ	Headquarters
HRH	His/Her Royal Highness
IC	Intelligence Corps
ID	Identity
IRA	Irish Republican Army
JARIC	Joint Air Reconnaissance Intelligence Cell
KCB	Knight Commander of the Order of the Bath.
KGB	*Komitet gosudarstvennoy bezopasnosti* (Committee for State Security)
LRDG	Long Range Desert Group

LSGC	Long Service and Good Conduct Medal
MAS	Manual of Army Security
MBE	Member of the British Empire
MC	Military Cross
MID	Mentioned in Despatches
MI5	Military Intelligence 5 (Security Service)
MI6	Military Intelligence 6 (Security Intelligence Service)
MM	Military Medal
MIO	Military Intelligence Officer
MSI	Military Security Instructions
NATO	North Atlantic Treaty Organisation
NBC	Nuclear, Biological and Chemical
NCO	Non Commissioned Officer. In British Army from lance-corporal to staff sergeant equivalent
OBE	Order of the British Empire
OM	Order of Merit
PI	Photographic Interpretation/Interpreter
Psy Ops	Psychological Operations
PWE	Psychological Warfare Executive
RA	Royal Artillery; Royal Academy
RAF	Royal Air Force
RFC	Royal Flying Corps
RAMC	Royal Army Medical Corps
RPG	Rocket Propelled Grenade
SD	*Sicherheitsdienst* ((Nazi Party) State Security)
SS	*Schutzstaffel* (Protection Squadron)
SOE	Special Operations Executive
SOXMIS	Soviet Commanders'-in-Chief Mission in West Germany.
TA	Territorial Army
UN	United Nations
Y Service	First and Second World Wars Intercept and Direction-Finding Signals Intelligence

Military Intelligence

Knowledge gives strength to the arm

As an island nation, the English regularly deterred or defeated internal and external threats, such as Spanish subversion during the Elizabethan era and Dutch naval incursions in the seventeenth century. When the Union was formed, that success continued in the expeditionary wars against France that ended with victory at the Battle of Waterloo. The country's most effective trench, the English Channel, has always been dominated by its guardian, the Royal Navy, and consequently the most dangerous threats not infrequently originated via the back door – from Ireland, and later from the air.

John Churchill, the first Duke of Marlborough, stated in the early 1700s that 'No war can be conducted successfully without early and good intelligence'. Briefly, military intelligence can be defined as the interpretation of information relating to the military, economic, political, sociological, and industrial infrastructure of regions, countries and organisations that pose, or may pose, a threat to the national interest. The interpretation is exercised through a sequence of mutually supporting processes usually known as the Intelligence Cycle and deciding the 'essential elements of information'. Human intelligence is the oldest intelligence source. Document exploitation ranges from the media to battlefield detritus. Signals intelligence is airwave interception. Imagery intelligence decodes. Technical intelligence investigates equipment of any size or type. Information is processed through analysis, interpretation and evaluated against other information to indicate credibility and then distributed without bias in time to be of use.

Imperial disasters after Waterloo of intelligence failures in Afghanistan, Crimea, the Indian Mutiny and defeats by assegai-wielding Zulus and rough-riding Boer farmers convinced forward-thinking Victorian officers that intelligence as a military asset in war and peace. During the Second Boer War (1899–1902), military intelligence developed into the three-legged stool of information collection, tactical Field Intelligence Departments and headquarters intelligence staff officers advising commanders. This led in 1904 to the War Office establishing the General Staff from brigade upwards to assist commanders in transmitting policies and orders. Features included forming the Military Intelligence Department within the Mobilisation Department and the raising of a hostilities-only Intelligence Corps of intelligence officers and a counter-intelligence cohort. Meanwhile, open and clandestine intelligence collection was being exercised through consuls, military attaches and officers on leave. In 1909, the General Staff formed the defensive MI5 (Secret Service Bureau) and offensive MI6 (Secret Intelligence Service Bureau).

First World War

When war broke out on 4 August 1914, the British Expeditionary Force (BEF) Intelligence Corps mobilised from a mix of fifty-five army officers and civilians, with mainly linguistic skills, and the Intelligence Police of 24 Metropolitan Police Special Branch officers for counter-intelligence operations in the rear areas. They became known as the Intelligence Police. The Corps was administered by the 10th (Service) Battalion, Royal Fusiliers, based in Hounslow. During the Retreat from Mons, Second Lieutenant Rolleston West earned its first DSO when he helped a Royal Engineer officer demolish a bridge.

As the Western Front stabilised into trench warfare, prisoners of war and refugees were processed through in an effective information collection. When the Germans intercepted the front-line telephone network, the British duplicated the apparatus in 1916 with the Intelligence Telephone, commonly known as I-Toc, which was operated by detachments of two Royal Engineer signallers and an Intelligence Corps NCO linguist, monitoring German telephone traffic in the front line. The Intelligence Police played a crucial security and counter-intelligence role in minimising espionage, sabotage and subversion.

The first Commandant of the Royal Flying Corps (RFC), Brigadier-General David Henderson, a former Director of Intelligence, realised the value of aircraft and balloons on the third flank of air as an intelligence asset and encouraged pilots and observers to spot and photograph enemy positions for interpretation by photographic analysts. Although this air reconnaissance induced aircraft 'dog fights', about 19,000 photographs were taken of the Somme battlefield between July and September 1916. As the demand grew, the School of Photography, Mapping and Reconnaissance was established in Farnborough. Throughout the war, Intelligence Corps officers posing as passport control in the Channel ports and embassies identified spies, talent-spotted potential agents and ran espionage circuits in the German-occupied countries. One of the most profitable was the collection of information on enemy troops being moved by railway.

Elsewhere, intelligence officers based in Egypt tackled subversion in Cyprus, which had recently been annexed from Turkey, and, using 'intelligence yachts', conducted clandestine intelligence operations along the coasts of Anatolia, Syria and Palestine. A fruitful network of young Jews run by Captain Lewen Weldon proved particularly valuable for the preparations by General Edmund Allenby in 1917 to defeat the Turks. Previously, the Intelligence Department in Cairo ignored their intelligence, which led to defeats at Gaza. The network was destroyed when the pigeon used to carry messages was distracted by a liaison in the loft of the Turkish governor in Haifa and was captured. This led to Turkish counter-intelligence destroying the network. In Mesopotamia, intelligence mistakes led to Imperial surrender at Kut-al-Amara in 1916. In East Africa, the disastrous landing at Tanga in November 1914 led to the restructuring of the intelligence organisation. South African Army intelligence played a major role in the destruction of the German cruiser *Konigsberg,* hiding in a muddy estuary south of Dar es Salaam. As Imperial forces drove the Germans from modern Tanganyika, a medical officer wrote:

> Far ahead of the fighting troops are the Intelligence officers with their native scouts. These officers, for the most part, are men who have lived in the country, who know the native language and are familiar with the lie of the land from experience gained in past hunting trips.

Two Victoria Crosses were awarded to intelligence officers, namely Lieutenant-Colonel Charles Doughty-Wylie at Gallipoli and Captain Frederick Hotblack, who was seconded into the Intelligence Corps in 1914.

After the Armistice in 1918, the Intelligence Corps was involved in the occupation of Germany until the late 1920s in a counter-intelligence role against subversion. Several progressive officers, notably Major (later Field Marshal) Gerald Templer, recognised the emerging threat from Nazi Germany and, when the BEF mobilisation plans were revised, it included an Intelligence Corps. The disbanded Intelligence Police reformed as the Field Security Police (FSP) in 1937.

Second World War

After defeat at Dunkirk, the Intelligence Corps was promulgated as part of the Army on 19 July 1940. The FSP reformed as Field Security (FS). The first FS course for Auxiliary Territorial Service (ATS) women was held in the same year. FS sections were widely dispersed and Travel Control Security sections covered all UK points of entry in harbours and airports to vet arrivals. Suspects were transferred to MI5. Among FS sections formed in the Far East, some were composite, with Indian, Burmese and African soldiers. Intelligence Corps officers with linguistic skills were employed at tactical interrogation centres and high-level Combined Services Detailed Interrogation Centres. Intelligence Corps linguists captive in prison camps proved invaluable. Photographic

Interpreters (PI) of mainly officers supported a wide range of operations, including Operation *Bodyline/Crossbow* identifying German V-weapon sites at Peenemünde and the Pas de Calais, and for the D-Day landings. Signals Intelligence had increased exponentially with Intelligence Corps linguists and analysts in divisional intercept units. About 400 Intelligence Corps men and women were associated with Bletchley Park and its equivalents in Jerusalem and Delhi supporting Middle East and Far East operations respectively. About 450 Intelligence Corps served with the Special Operations Executive (SOE) as agents and instructors. Three FS sections helped assess the suitability of agents under training. By January 1945, the Corps numbered about 3,050 officers and 5,930 soldiers against 2,000 in 1939. Colonel James Ewart interpreted for General Montgomery during the German surrender at Luneburg Heath in May 1945.

Post-1945

After Germany and Japan surrendered, the Intelligence Corps played an important role in the occupation of Germany, Italy and Japan, and in liberated territories. Keen to reduce costs, the War Office attempted to disband the Intelligence Corps; however, global instability and the Cold War saw the Corps deployed in all the counter-insurgency operations and wars fought to protect British interests. The murder of two FS sergeants during the Palestine Emergency (1945–48) was a factor in the British ending its Mandate, and, in the Kenya Emergency (1953–56), the Air Photographic Interpretation Unit provided intelligence for bombing the forest occupied by the Mau Mau.

In the 1950s, three important decisions were made in response to the increasing threat from the Soviet Union. First, to increase the efficiency of military intelligence by,

in 1957, transferring 100 Regular officers into the Intelligence Corps – the 'First 100'. Previously, talented individuals were recruited as intelligence officers but, since there was no career path in the skill, they returned to civilian life. Secondly, the finishing of National Service in 1960 led to the Intelligence Corps taking responsibility for providing intelligence sections from brigades to the War Office and with national and international military organisations. Previously, intelligence sections were assembled from units in formations. Associated with the decision was progressive training in intelligence, and security for other ranks and the wider military and naval community. The third decision was in response to the spread of communism orchestrated by hostile intelligence services and involved forming a Security Branch within the Directorate of Military Intelligence. Its role was to manage the protection of the Army and its civilian components from espionage, subversion and sabotage, and, later, international and domestic terrorism. The expertise the Intelligence Corps had gained in Field Security since 1914 was exploited by it taking the lead in Protective Security and Counter-Intelligence. The arrival of the First 100 saw the Corps develop a regimental structure of Intelligence and Security Groups. For instance, Intelligence & Security Group (Germany) consisted of two security companies supporting 1st British Corps and the Rear Areas, a Photographic Interpretation Company, a Signals Intelligence company embedded with a Royal Signals regiment and an Intelligence company. The Intelligence Corps (TA), which has played a crucial reserve role, was also formed into a Group.

The new organisation was first tested during the Indonesian Confrontation (1962–66). During the campaign, Military Intelligence Officers (MIO) liaising with Special Branch were reinforced by Intelligence Corps Field Intelligence NCOs (FINCOs), a practice that was extended to the garrisons in Belize, Cyprus, Hong Kong, Singapore and Northern Ireland. In South Arabia (1964–66), the wife of an Intelligence Corps officer was murdered by a terrorist bomb. In Operation *Banner* in Northern

Ireland, Continuity Junior NCOs were deployed to Security Forces' forward operating bases to manage local intelligence. The Falklands Campaign (1982) was pivotal because, for the first time, the senior intelligence officer at HQ Land Forces Falkland Islands was an Intelligence Corps officer. The internationalisation of operations since 1990 in Kuwait, the Balkans, Iraq and Afghanistan saw the Intelligence Corps reform into Military Intelligence Brigades to meet the increasingly complex intelligence requirement demanded by allies and partners.

A crucial element in the success of intelligence officers and the Intelligence Corps is that any rank can be sent anywhere at short notice, often alone, often in civilian clothes, and for that soldier to use initiative and experience to achieve the objectives of making sensible security recommendations and predicting accurate intelligence.

In July 1977, the Intelligence Corps adopted Cypress Green, the traditional colour of military intelligence, as the colour of its beret and three years later was declared to be an 'Arm' in recognition of its role to be in close in combat with the enemy – that is, hostile intelligence services. By now, the Corps was directly recruiting women. Homes of the Intelligence Corps have included Winchester and Wentworth Wodehouse during the Second World War and then Maresfield, Sussex, Ashford, Kent and, currently, Chicksands, Bedfordshire. Regimental Alliances include those with the Australian Intelligence Corps and the Intelligence Branch of the Canadian Armed Forces, and Bonds of Friendship with the Malaysian Intelligence Corps and United States Military Intelligence Corps. The Intelligence Corps is honoured to have Field Marshal HRH Prince Philip, Duke of Edinburgh, KG, KT as its Colonel-in-Chief.

The Military Intelligence Museum

The Intelligence Corps Museum was first established at Maresfield in 1947; however, the nature of military intelligence and security has meant that it suffers from a lack of public visibility because it has been confined within barracks with its size largely restricted to a relatively small space.

When the Corps relocated to Templer Barracks, Ashford, Kent, in 1966, four years later, it was moved into a purpose-built building opposite the Guardroom and was opened by Field Marshal Templer, who had extensive intelligence experience and had been a supporter of the Corps for many years. After the Intelligence Corps moved to Chicksands, Bedfordshire, the Museum was re-opened in July 2005. Parallel to presenting a high quality object display, interactive elements included 'In the name of the Rose', commemorating Intelligence Corps who lost their lives while serving, and 'Honours and Awards', listing those awarded gallantry and commendation medals, complete with citations. Generous donations by a local benefactor, Mr Julian Barnard, has allowed a world-class display on the SOE. In the same building is the Medmenham Collection, relating to Imagery Intelligence. In 2007, the Intelligence Corps Museum was incorporated into the Military Intelligence Museum in order to reflect the nature of intelligence and security as a Defence asset. The Friends of the Intelligence Corps Museum give considerable financial support.

Pre-1914

The Amiens Medal

The Treaty of Amiens between Britain and France after twenty years of war was commemorated in 1801 by the Amiens Medal, inscribed as 'Preliminaries of Peace Between Great Britain and France; Signed October 1st 1801'. Although the status quo established by Napoleon in Europe was accepted, he was regarded as a threat to British interests and, in May 1803, Great Britain engineered the renewal of hostilities that eventually ended at the Battle of Waterloo on 18 June 1815.

Waterloo Medal awarded to Lieutenant-Colonel Colquhoun Grant (1780–1829)

Grant was commissioned in 1795. When, in 1811, he joined the Peninsula Army fighting the French in Spain, General Arthur Wellesley (later Duke of Wellington) appointed him as his 'Exploring Officer' in the specialist Peninsula Corps of Guides. Grant was captured collecting intelligence from behind French lines in 1812 and accepted parole. His servant was less fortunate and was shot. When the French discovered he was passing messages to Wellesley and sent him to Paris for interrogation, he saw a letter written by the French commander in Spain that he was not to be exchanged for a prisoner of equal rank. Grant therefore escaped and remained in Paris and, masquerading as an American officer, passed information to Wellesley, then advancing into France. Rejoining the Peninsula Army, Wellesley appointed him Head of Intelligence and, during the 1815 Waterloo Campaign, formed an Intelligence Department for the

first time in British military history, with Grant as its head. During the French advance into Belgium, one of Grant's intelligence couriers was detained by a Hanoverian brigadier for twenty-four hours, which delayed Wellington from deploying his army by a day. Grant later commanded a brigade in the First Anglo-Burmese War.

Private J. N. Richardson

Private Richardson served as a farrier with 3rd Company, Commissariat and Transport Corps during the Egyptian Campaign (1882–89). While serving in General Sir Garnet Wolseley's column, he was awarded the Khedive's Bronze Star for 'intelligence reconnaissance to find out dispositions and strength of enemy forces of Arabi Pasha'. Richardson constructed the display and used his campaign stirrups.

Adventures of a Spy by Lieutenant-General Robert Baden-Powell OM GCMG GCVO KCB DL (1857–1941)

Plan of a Dalmation fort incorporated into a butterfly sketch;
Fleur de Lis
Field Intelligence: Its Principles and Practices (1904) by Lieutenant-Colonel David Henderson.

Baden-Powell was commissioned in 1876 and, while an intelligence officer in Malta, visited several countries in which he incorporated sketches of military installations into his drawings of nature. In 1884, he published *Reconnaissance and Scouting* and he ran a course for reconnaissance while while commanding the 5th Dragoon Guards in India. Those who passed were awarded with the Fleurs de Lis badge, now associated with the Scout movement. After the Second Boer War (1899–1901), he was appointed Inspector-General of the Cavalry and insisted that a major role of cavalry regiments was reconnaissance. This factor gave the British a significant advantage in the First World War. Meanwhile, the Scout movement had grown and Baden-Powell's connection with it occupied him for the rest of his life.

Lieutenant-General Henderson showing his RFC wings

Written from his experience as Director of Military Intelligence during the Second Boer War, *Field Intelligence: Its Principles and Practices* established Henderson as an innovator of modern intelligence. Divided into eight sections, the book covers 'Personnel and Organisation', in which he recommends forming an Intelligence Corps; the 'Acquisition of Information' from reconnaissance, prisoners, documents and the Secret Service; the Value of Incomplete Intelligence'; the 'Guiding of Troops'; 'The Frustration of the Enemy's Endeavours to Gain Information'; and the distribution of intelligence using maps and reports. In 1907, Henderson wrote *The Art of Reconnaissance*. In 1911, aged forty-nine years, he learnt to fly, and helped establish the RFC, before commanding it for most of the First World War. He was instrumental in reforming it into the RAF.

Field Intelligence Detachment (FID) Badge

During the Boer War, Henderson formed Field Intelligence Departments for each column. These became the foundations on which the Intelligence Corps was later formed.

Periscope

One of the earliest aids to Imagery Intelligence was the periscope. When the lens is moved, it brings the two photographs into stereo. The narrative of the reverse of the photographs states 'Behind the Orange River entrenchments holding the bravely advancing Boers' and is dated January 1900. At the time, the periscope was a new innovation and much in demand in high society drawing rooms.

Medals of Major C. D. D. Shute (West Surreys)

During the Second Boer War, Major Shute served as Head of Intelligence in General Frederick's advance to relieve Mafeking in May 1900. The medals are his Queen's South Africa Medal with Cape Colony and Transvaal clasps and the King's South Africa Medal (1901/02).

Medals of Conductor David Scotland (South African Army Service Corps)

Conductor Scotland served with Kitchener's Fighting Scouts during the Second Boer War. The medals are 1899 South Africa with clasps for Transvaal, Orange Free State and Cape Colony, 1914–1918 Star, 1914 British War Medal and 1918 Victory Medal. The South Africa cap badge is inscribed 'Union in Strength' and *'Eendracht Maakt Macht'*.

Natal Rebellion Medal

The 1906 Natal Rebellion Medal was awarded to Trooper Friend for service to quell a Zulu revolt. The medal is inscribed as *'Tpr L Friend, Intelligence Corps'* at a time when no such organisation existed in the British Army.

First World War

Intelligence Brassard

When the Intelligence Corps was formed in August 1914, those seconded to it wore grey brassards inscribed 'IC' in green. Note the Fleurs de Lis, indicating that the wearer had passed the scouting course.

Jacket worn by Lance-Corporal Castle

Frank Castle (17th Lancers) was seconded to the Intelligence Corps, as identified by the brassard. The 1902 Pattern Service Dress tunic is worsted. The breast pockets were for personal items and the AB64 Pay Book, which gave personal details. An internal pocket under the lower right flap was used to store the First Field Dressing. The brass buttons are annotated with the 'Skull and Crossbones' insignia of the 17th Lancers. Long Service and Good Conduct stripes were placed on the lower sleeves.

First World War Service Dress Officer's Tunic

The tunic was owned by Lieutenant E. R. A. J. Gout. Whereas other ranks were issued with uniforms, officers were given an allowance. The jacket was tailored by F.W. Flight, Military Outfitter of Winchester, Aldershot and London. Uniforms were usually quality Barathea wool and fully lined with rank incorporated into the lower sleeve. The green collar gorgets shows it was worn by an intelligence officer. Note the badges of rank are on the shoulder epaulettes, as opposed to the lower sleeve.

Intelligence Officer's Cap

The cap was owned by Lieutenant E. R. A. J. Gout. The bottle green band around the cap shows that it was used by an intelligence officer. The cap badge is General Service Corps, which was the holding unit for officers with specialist skills.

Lieutenant-Colonel Stanley Woolrych OBE

Stanley Woolrych was commissioned into the BEF Intelligence Corps in 1914 and by 1916 was masquerading as a Passport Control Officer in Paris, running an espionage network in German-occupied France and Belgium. One major source of information was military deployments by railway. Prior to the Second World War, he ran the Field Security Police school at Mychett and during most of the war was Chief Instructor at the SOE Finishing School at Beaulieu. The medal array is the OBE, 1914–18 Star,

British War Medal, Victory Medal with MID Oakleaf, 1939–45 Defence Medal. Lower are the Belgian Order of the Crown and Croix de Guerre awards.

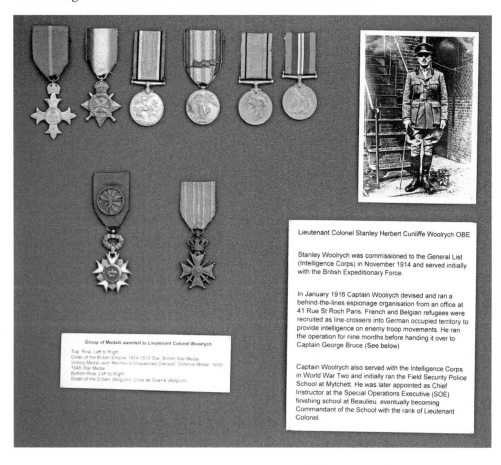

Group of Medals awarded to Lieutenant Colonel Woolrych

Top Row, Left to Right:
Order of the British Empire, 1914-1915 Star, British War Medal,
Victory Medal (with Mention in Dispatches Oakleaf), Defence Medal, 1939-
1945 War Medal
Bottom Row, Left to Right:
Order of the Crown (Belgium), Croix de Guerre (Belgium)

Lieutenant Colonel Stanley Herbert Cunliffe Woolrych OBE

Stanley Woolrych was commissioned to the General List (Intelligence Corps) in November 1914 and served initially with the British Expeditionary Force.

In January 1916 Captain Woolrych devised and ran a behind-the-lines espionage organisation from an office at 41 Rue St Roch Paris. French and Belgian refugees were recruited as line-crossers into German occupied territory to provide intelligence on enemy troop movements. He ran the operation for nine months before handing it over to Captain George Bruce (See below)

Captain Woolrych also served with the Intelligence Corps in World War Two and initially ran the Field Security Police School at Mytchett. He was later appointed as Chief Instructor at the Special Operations Executive (SOE) finishing school at Beaulieu, eventually becoming Commandant of the School with the rank of Lieutenant Colonel.

Portable Pigeon Coop

Pigeons were used during both world wars for intelligence purposes, principally carrying messages from agents. Pigeons were smuggled by couriers and later dropped by parachute. During the Second World War, a valued MI14 (Germany) source was homing pigeons returning to England.

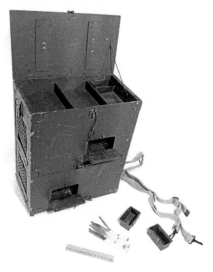

Trench Periscope

A Lifeguard Periscope manufactured by F. Duerr & Sons of Manchester. The extendable periscope was an innovation that enabled observers to view the enemy from 50 cm (20 inches) below the parapet up to a range of 91 m (100 yards).

'The Interrogation'

The 1946 Manual of Military Intelligence Part 7 (PW Intelligence) defines that the object of interrogating prisoners of war and civilians is to acquire information that, together with other intelligence, enables commanders to assess the enemy threat. It depicts an incident a week before the last German offensive in March 1918, initially sketched by the artist, Francis Dodd, on toilet paper. It depicts a British captain from GHQ and an Intelligence Corps lieutenant interrogating a German from the 13th Infantry Regiment. Notably, the Intelligence Corps lieutenant is middle-aged. In 1916 Dodd was appointed as the official artist with the War Propaganda Bureau. A copy has been a feature of the Military Intelligence Museum and its predecessors for several decades.

'I Won't!'

Equally important is that prisoners resist interrogation. The 1929 Third Geneva Convention (Treatment of Prisoners of War) was adopted as a direct result of mistreatment during the First World War. The painting by Fortunio Matania depicts a dishevelled Highlander refusing an invitation from German officers to sit in a chair.

Medals of Brigadier-General John Charteris CMG DSO

Brigadier-General Charteris was General Douglas Haig's Chief of Intelligence at First Army and the BEF when Haig was in command. When Charteris returned to England in 1916 on sick leave, Major James Marshall-Cavendish (Intelligence Corps) was sent to Army HQ and discovered that the intelligence briefings being given by Charteris were overly optimistic and, therefore, inaccurate. Medals left to right are Companion of the Order of St George; DSO; 1914–1915 Star; War Medal 1914–18; Victory Medal 1914–18; Delhi Durbar Medal, Belgian Order of the Crown, Croix de Guerre (Belgium); Chevalier Legion d'Honneur (France); Distinguished Service Medal (US) and Order of the Rising Sun (Japan).

Bust of General Sir James Marshall-Cornwall KCB CBE DSO MC

Marshall-Cornwall was commissioned into the Royal Field Artillery in 1907. Fluent in several languages, when the First World War broke out he was seconded to the Intelligence Corps and regularly crawled into No Man's Land to listen to German conversations. In 1918, he was posted to MI3 (Europe) and held diplomatic appointments between the two world wars. In 1941, Churchill sent him on a fruitless mission to persuade Turkey to enter the war on the Allied side. In 1943, he was dismissed from commanding Western Division in a political wrangle about the safety of Liverpool Docks and spent the rest of the war promoting better relations between SOE and MI6. Between 1947 and 1951, he edited captured German documents for the Foreign Office.

Sergeant Albert Long MM and Bar

Long enlisted into the West Yorkshire Regiment in 1914 and was seconded to the Intelligence Corps. Deployed to front-line observation posts to identify targets for mortars, artillery and bombing, he was awarded the MM and Bar for acts of gallantry on 19 March and 13 September 1918. His medals include the British Victory Medal and British War Medal.

Gas Hood

After Germany launched the first major chemical weapons attack on the Western Front in 1915, the British issued simple gas masks of cotton mouth and nose pads soaked in chemicals. They were replaced by the P (Phenate) Helmet made of layers of flannel dipped in sodium phenolate and glycerine that gave protection against chlorine and phosgene, but not tear gas. Uncomfortable and prone to causing eye and skin irritation, the hood was replaced in 1916 by the Small Box Respirator. While death rates from gas were low, those worst affected experienced lifelong discomfort. Gas attack remains a psychological weapon.

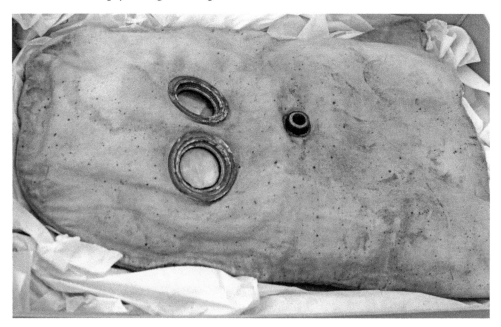

Personal Medical Kit

A pack purchased by an intelligence officer before departing for France in 1914. The small tubes contain Dover Powder to treat colds and fever; boric acid to keep feet healthy; aspirin; quinine against fever and malaria; potassium to purify water and cascara laxative. Also included are a small bandage, an arm sling, a box of elastoplasts and dental floss. Standard issue field dressings consisted of two bandages each with a dressing wrapped in a waterproof cover.

Inflatable pillow used by Sir Oswald Hornby Joseph Birley MC RA

An inflatable pillow purchased by Second Lieutenant Birley and marked '10th (Intelligence Corps) Royal Fusiliers'. A noted painter, he was involved in air photographic interpretation and was awarded the MC.

Example of a Leave Pass

A Leave Pass issued to 6147 CSM Henderson G. A. of 3rd Battalion, AIF (Australian Imperial Force), valid from midday 18 December 1917 to midnight 28 December 1917 in Poole, Dorset. It was issued by the Orderly Room, HQ AIF Depots in UK in Bhuripore Barracks, Tidworth.

German 'Pikelhaube' Helmet

A leather German 'pikelhaube' helmet, inscribed '*Mit Gott fur Koenig und Vaterland*' ('God, King and Fatherland') and 'FR' (*Friedrich Rex*). Manufactured from leather, it was no more suitable for trench warfare than the caps first worn by the British. In 1916, the Germans replaced it with the iconic '*Stalheim*' coal scuttle helmet, which significantly decreased head wound fatalities. In 1915, Mr John Brodie designed the British 'pudding bowl' Helmet, Steel, Mark I.

Inter-War 1918 to 1939

Field Security Police and their Employment on Military Intelligence (B) Duties (1923)

By 1923, the Intelligence Police had been renamed the Field Security Police (FSP). Issued as a secret supplement to the Manual of Military Intelligence by the General Staff, the manual describes the FSP as an executive element of the Defence Security Intelligence Section (MI5) in the field with 'powers as soldiers and as constables'. Topics covered include compiling White Lists (friendly sources) and Black Lists (known and suspect enemy agents); wearing civilian clothes; Travel Control Security at ports and airports and the close relationship with the Provost Marshal. In some respects, this document laid the foundations of the role of the Intelligence Corps.

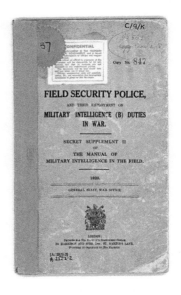

Counter-Intelligence and Protective Security – German Security Intelligence (1924)

Subtitled a *Guide to the Investigation and Prosecution of Espionage and High Treason* from the 1914 German Military Secrets Act, the document was translated by the British occupying powers and detailed the evidence that would be needed after the discovery of a traitor through surveillance, detection, arrest and trial. It includes methods to intercept correspondence and the 'Reconstruction and Deciphering of Postal Matter'.

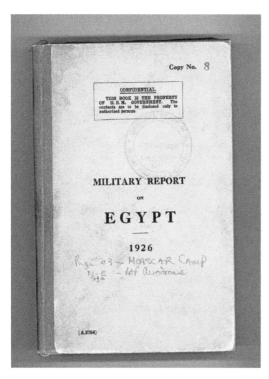

Military Intelligence Report — Egypt (1928)

Collected from a variety of sources including official reports and traveller and media observations, hostile and country reports are essential basic intelligence and regularly updated. Subjects included geography, infrastructure, political and social structure and armed forces. The British had first arrived in Egypt in 1882 and consequently this report is detailed. Such assessments are of particular value in places not frequented by the British. When the 1982 Falklands War broke out, little information was available for the Task Force.

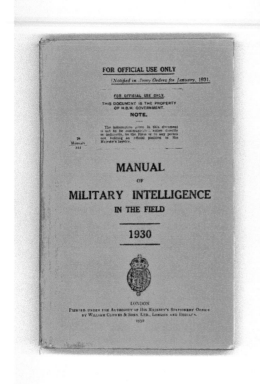

Manual of Military Intelligence in the Field (1930)

This wide-ranging manual describes general principles of Military Intelligence, covering reconnaissance, ground/air coordination, prisoner, civilian and document intelligence, military survey, characteristics of intelligence in underdeveloped countries and the management of intelligence records. The chapter titled 'Security Intelligence in the Field' includes conduct after capture, censorship and propaganda. The intelligence structure from GHQs to unit level and the Intelligence Corps is described, which existed in name only. Appendices included a specimen syllabus for command, brigade and unit intelligence section courses.

Second World War

Field Security Police (FSP) Brassard

Field Security Police (FSP) Cap Badge

To retain anonymity, when the FSP was formed in 1937 the soldiers wore the Corps of Military Police cap badge; however, some objected to being associated with the military police and filed its nomenclature from the badge.

Medals of Brigadier W. F. Jefferies CBE DSO (Overleaf)

Commissioned into the Royal Dublin Fusiliers in 1914, William Jefferies was wounded twice. Mentioned in Despatches and awarded the DSO during the First World War, he was deployed to Ireland during the Irish War of Independence (1919–21) and was possibly a member of the Cairo Gang of British intelligence officers attacked by the IRA. In 1940, he was instructed to form the Intelligence Corps at Pembroke College, Oxford University, and was the first Colonel Commandant. Between 1943 and 1945, he was Deputy Head of Psychological Warfare in the Middle East and Central Mediterranean.

Intelligence Corps Cap Badge

When on 19 July 1940 King George VI approved the inclusion of the Intelligence Corps as part of the British Army, the new Corps took as its badge a double Tudor Rose (representing secrecy), flanked by laurel leaves (victory) resting on a scroll inscribed 'Intelligence Corps' and topped by the Monarch's crown (loyalty).

Censorship Department Mugs

Censorship Department Badge

The First World War Censorship Department badge may have provided the model for the Intelligence Corps cap badge. Its inscription in Greek translates as 'In Secrecy We Work'.

Field Security Section (FSS) Brassard

Field Security Section Counter-Intelligence (CI) Brassard

The FSS brassard was worn by those serving in FS section. The exhibit is thought to be homemade, probably in a remote place, such as Persia and Burma. Soldiers were issued a small holdall known as a 'housewife', containing a thimble, grey darning wool, linen thread, needles, brass and plastic buttons. The issue was discontinued in about 1971. Some detachments adopted the nomenclature 'CI' (Counter-Intelligence).

Portrait of an unidentified ATS attached Intelligence Corps

ATS Jacket

The ATS jacket was worn by Sergeant Kay Norman (Intelligence Corps), who was attached to the Beaumanor Hall Y Service Station intercepting German Army wireless transmissions. Analysis was sent by motorcycle despatch to Bletchley Park. Above the upper left pocket is her Intelligence Corps badge.

Periodical Notes on the German Army

Prepared by MI14 (Germany), the *Periodical Notes on the German Army* was published in March 1940 and replaced *Tactical and Technical Notes on the German Army*. No. 30 (August 1940) listed lessons learnt from the Battle of France and possible German tactics for an attack on Great Britain. No. 38 (20 March 1943) analysed the German XI Air Corps airborne assault on Crete. In modern terms, the Periodical Notes were the equivalent of Supplementary Intelligence Reports.

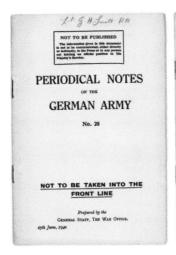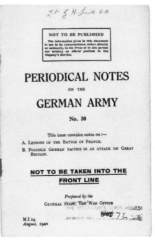

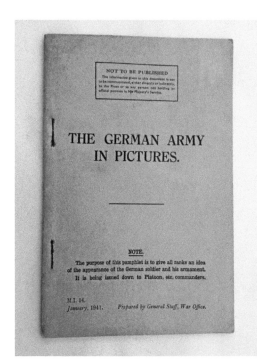

The German Army in Pictures (1941)

The document was a recognition aide showing black and white photographs of field uniforms, including airborne and motorcycle troops, supported by basic specifications of armoured cars, tanks, small arms, infantry support weapons and pneumatic assault boats.

Enemy Weapons – Part 1 – German Infantry Weapons (November 1941)

Published after the battles for France, Greece and Crete and during the fighting in North Africa, the pamphlet is essentially Technical Intelligence and was designed to help troops use captured machine-carbines (modern submachine guns), machine guns, anti-tank rifles, carbines and rifles and light mortars. It gave identification criteria for small arms ammunition and supplied conversion tables for distance and angles.

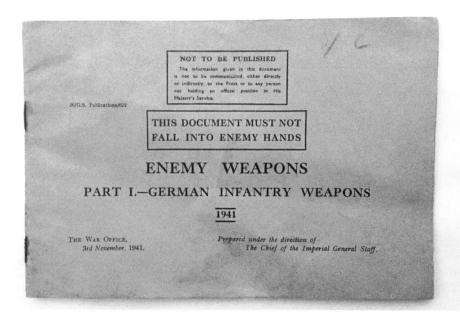

Pocket Book of the German Army (1943)
Regimental Officer's Handbook of the German Army (1943)
The Pocket Book superseded *New Notes on the German Army No. 1 – Armoured and Motorised Divisions* (1942) with material gained from the fighting in North Africa and Russia. The Intelligence Officer's Handbook gave guidance on German tactics, unit organisation, weapons and uniforms and running a battalion-level intelligence section.

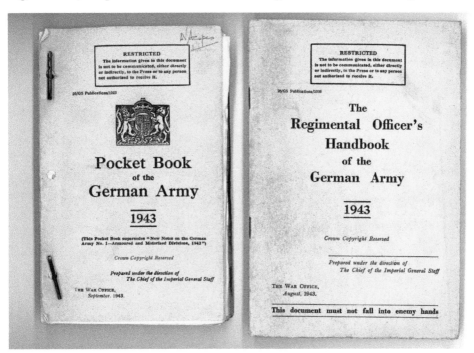

Enigma Cipher Machine
The Enigma electro-mechanical rotor cipher machine was developed by Germany to protect commercial, diplomatic and military communication. Polish Military Intelligence broke the codes in 1932 and they passed an Enigma to the British and French in 1939. Equally important for Signals Intelligence were the enemy wireless procedures, codes and manuals captured on the battlefield, at sea from U-boats and from aircraft.

Desert Signals Intelligence Scenario

The scenario depicts an Intelligence Corps analyst using a US National HRO-5 receiver to intercept enemy battlefield transmissions. The wireless was valve-based and received Short Wave Band and Carrier Wave (Morse). On the front control are two key features: a precision tuning dial and pluggable coils stored in a wooden locker that enabled the receiver to intercept thirteen frequency spreads over four bands. The name HRO is the abbreviation of 'Hell'va Rush Order' (Hell of a Rush Order)!

Medals of Captain John Makower MC MBE MID Bronze Star (US)

For sixteen months, Captain Makower (Intelligence Corps) commanded the Intelligence Detachment of 101st Special Wireless Section when it was attached to the 4th (Indian) Division during the intense fighting in North Africa, in which the understaffed detachment worked abnormally long hours, often under fire. Makower showed the inspirational leadership that laid the foundations for intercept units in the field. **Sergeant William Swain** (Intelligence Corps) was in the Section between July and December 1941. Disregarding personal safety and comfort, he helped convert an experimental unit to one of significant operational importance. They were awarded the MC and MM respectively. A painting of them by Terence Cuneo hangs in the Officers' Mess.

George Medal of Corporal Frank Thomas Turner Fowler

Lieutenant Allan MacDonald and Corporal Fowler, both Intelligence Corps, were serving with 118 Special Wireless Section supporting US Army Air Force operations at Great Glenham when a B-17 Flying Fortress with a full bomb load crashed. When MacDonald called for volunteers, Fowler responded and, between them, they rescued several crew members but were killed when a bomb exploded. Both were awarded George Medals.

Clayton Collection

After serving as a Royal Engineer in the First World War, Patrick Clayton was employed by the Egyptian Survey Department to map the North African desert. Italian cartographers were doing the same. When the Second World War broke out, Clayton was commissioned into the Intelligence Corps and, with Ralph Bagnold, another desert explorer, formed the LRGD to conduct reconnaissance patrols behind enemy lines and guide raiding units to their targets. The LRDG cap badge depicts a scorpion within a wheel. The opposing patrols agreed to cache water and petrol, and, to prevent tribesmen scavenging them, each was disguised as a military grave with a cross topped by a steel helmet. A fragile exhibit of a cache deposited by a German patrol for two agents infiltrating into in Cairo in 1942 is in the Museum. But their lifestyle attracted attention, and both men were arrested by 259 FSS in July 1942, an episode dramatised in the film *Foxhole in Cairo*. The cache was found in 2000 by the son of Clayton, Lieutenant-Colonel Peter Clayton, also a desert explorer, who presented it to the Museum. In January 1941, Clayton was wounded and captured in a clash with the Italian Auto-Saharan Co. and sent to a German prison camp, where he used his survey skills to make maps and forge German document stamps and maps.

Lieutenant-Colonel Patrick Andrew
Clayton DSO MBE

Escape Map and Forged German
Document Made by Clayton as a
Prisoner

Medals

Lance-Corporal Anthony Coulthard

Coulthard was an Oxford University graduate in modern languages, who had travelled and studied in Germany. In May 1940, he was captured near Amiens while serving with 32 (45th Division) FSP and was sent to Stalag XXA near Torun in Poland where he became the Camp Interpreter. When Sergeant Foster, who had been captured in Norway, suggested they should escape, Coulthard taught him German; however Foster's accent was poor, and he adopted the persona of being Hungarian. On 23 August 1942, they escaped and, masquerading as representatives of Siemens, travelled in some style by train to Munich and eventually reached the Swiss border at Lindau. Coulthard was allowed to cross but when he saw that Foster was being questioned by the frontier policeman, he returned to help and it was then that their papers were exposed as forgeries. Coulthard escaped eight more times. In mid-January 1945, in the grip of a particularly severe winter, the Germans marched 100,000 Allied and Russian prisoners in Poland and eastern Germany west away from the advancing Russians in the Long March. However, hundreds died from the privations of captivity. Those unable to keep up were often shot. Weakened by dysentery, tuberculosis and exposure after being forced to wash in the River Elbe, Coulthard died on 24 March 1945, aged twenty-six

Lance-Corporal Coulthard, Intelligence Corps, (far right with spectacles) as a prisoner of war

of Coulthard to Quickborn village cemetery. On 30 July 2015, Lance-Corporal Coulthard was re-interred in the Commonwealth War Graves Commission cemetery at Becklingen. The German officer in command of the column was later executed for war crimes. For his gallantry, Coulthard was awarded a posthumous Mention in Despatches.

Sergeant Frederick Foster, 1/8th Sherwood Foresters (TA) with members of 148 Infantry Brigade (TA)

Camp Escape Map

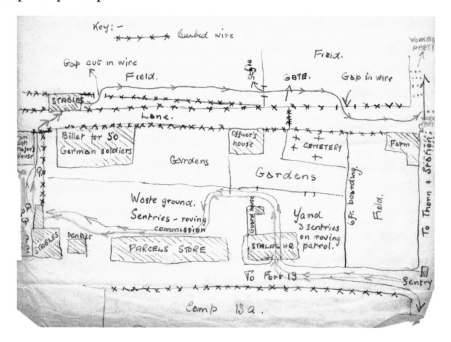

Part of the SOE display showing an agent unloading a canister

The Special Operations Executive (SOE) was a British Second World War organisation formed in July 1940 from within the Ministry of Economic Warfare by Hugh Dalton to conduct espionage, sabotage and reconnaissance in countries occupied and attacked by Germany, Italy and Japan, except in areas demarcated by the USA and Soviet Union. The organisation directly employed or controlled just over 13,000 people, about 3,200 of whom were women. The Intelligence Corps provided three FSS to SOE with a variety of roles.

Colonel Julius Hanau OBE

Hanau was a South African commissioned into the Army Service Corps in 1914. By 1918, he was Deputy Chief of Staff to the British Mission in Yugoslavia and, settling in Belgrade, he advised on Balkan political matters until 1939. In 1940, he was commissioned into the Intelligence Corps and led attempts to sabotage German imports from Romania by blocking the River Danube at its narrowest point. When this proved impossible, he resorted to sabotaging railway traffic until German counter-intelligence detected his activities and forced the Yugoslavs to expel him. Moving to Cairo, he ran the SOE Desk in Cairo until he died. The medal set also shows the 1918 Serbian White Eagle with Swords, Knight Commander of St Sava of Yugoslavia (2nd Class).

Turrall Collection

A geophysicist with an engineering degree from Cambridge University, Major Turrell was a Royal Engineer in the First World War. In 1939, he was commissioned into the Intelligence Corps and was awarded the MC while serving with the Sudan Frontier Force in Abyssinia. Recruited by the SOE Force 133 (Balkans), he took part in operations in Crete. By 1944, he was in the Far East serving as an Operations officer with Wingate's Special Force, known as the Chindits, then conducting the deep penetration Operation *Thursday* behind Japanese lines. Joining Force 136 (South-East Asia), aged fifty-four years, he twice parachuted 'blind' into Japanese-occupied Burma, organised reception parties for other agents and trained several hundred Levies. In April 1945, he led an attack on the Japanese *Kempei Tai* (military police) HQ and supply depot at Kyaukki, killing 550 of the enemy, and later directed air attacks and shelling on enemy positions east of the town, which accounted for a further 950 of the Japanese. On 16 August, the day Japan capitulated, he attempted to convince the 55th Division to surrender, but its officers were unaware of the surrender and handed him to the *Kempei Tai*, who treated him as a spy. Reports of his capture led to consternation at HQ 12th Army, and South-East Asia Command then tried to ensure controlled surrenders until a Japanese officer informed the Division that the war was over. The *Kempei Tai* released Turrall. Interestingly, he had undertaken a similar venture when the Armistice was declared on the Western Front in 1918.

Major Guy Turrall DSO MC

Tropical Dress Jacket

Above the medals are parachute wings. During the Second World War, some of those who completed parachute training but were not part of the Army Air Corps (airborne forces) wore their wings above the left upper pocket.

Battle Dress Jacket

Service Dress Jacket

Chindit badge and Japanese Cap

Medals of Major Frederick Vernon Webster, Intelligence Corps (India)

In 1942, Lieutenant-Colonel C. E. Gregory (Royal Garwhal Rifles), GHQ Head of Intelligence in Calcutta, formed Z Force (Burma) from expatriates who had worked in the Burmese forests and Chin, Kachin and Karen to collect intelligence during the dry seasons on Japanese activities. Major Herbert Casten and Captain Webster, an Anglo-Indian, formed a patrol and during the 1943 Chindit Operation *Longcloth* guided columns along the 'Secret Track' that bypassed Japanese garrisons into Burma. Initially, Z Force patrols from India walked to their operational areas; they were later dropped by parachute. In its three years of existence, Z Force was awarded one CBE, two DSOs, four OBEs, four MBEs, seventeen MCs with Bars to two, and sixteen Burma Gallantry Medals.

Le Chene Collection

Lieutenant Pierre Le Chene (Intelligence Corps), a bilingual French-speaking Briton, joined the SOE French Section. In early May 1942, he was parachuted into Central France as a wireless operator for a Resistance leader in Lyon during a period when German-inspired Vichy French counter-intelligence operations were disrupting the Resistance. Although HQ SOE did its best to forge accurate ID documents, they often contained errors. In common with other agents, Le Chene persuaded a friend employed in the office of the Mayor of Monte Carlo to produce a genuine ID card. When the Allies landed in North Africa in Operation *Torch* on 8 November and the tension escalated as the Germans prepared to occupy Vichy, eighty Direction-Finding vans moved into Lyon to pinpoint clandestine wirelesses. Next day, Le Chene was captured while transmitting to London after his look-out failed to warn him about a van. He was first British officer to be interrogated by the notorious SS Captain Klaus Barbie, the head of the Gestapo in Lyons. Barbie later admitted that Le Chene had betrayed no one. Le Chene was sent by Barbie to the SD in Paris where, in spite of brutal torture that left him with a limp and no nails in his fingers, he resisted by refusing to betray anyone. He later said, 'Each time, when they came at me and I knew what would happen, I told myself, stay with it just this once more.' After spending ten months in solitary in Fresnes Prison, he was transferred to Mauthausen concentration camp and sent to its brutal work sub-camp of Gusen. When US forces liberated the camp in late April 1945, Le Chene weighed six stone and was ill with typhoid.

Meanwhile US forces had found a box in Sachsenhausen concentration camp containing his property, including the Morse key, ID papers and a file of his interrogation in Lyons, and handed it to British Military Intelligence. It was offered to Le Chene in 1968 – much to his surprise.

The medals are the Officier de la Legion d'Honneur, Croix de Guerre avec Palms and several medals associated with the French Resistance and his deportation to Mauthausen.

Captain Pierre Louis Le Chene MBE after ten months' convalescence following his release from captivity in 1945

Morse Key and French ID Card

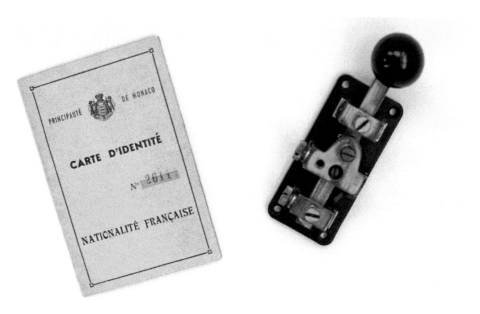

Medals

Legion d'Honneur DPV

The Order of the Légion d'Honneur is the highest decoration in France and is divided into five classes: Chevalier (Knight), Officier (Officer), Commandeur (Commander), Grand Officier (Grand Officer) and Grand Croix (Grand Cross). Those who display exceptional courage in which they put their lives at risk can be awarded the 'Décorés au Péril de leur Vie' ('Decorated at Peril of your Life'). Its award to a foreigner is rare. Le Chene is thought to be one of two Britons to receive the medal.

Neurath Medal

The Neurath medal – in fact, the Grand Cross of the Order of the German Eagle 1937 – belonged to Baron Konstantin von Neurath, an ambassador of the Third Reich, who lived in Ober Riexingen, about 20 miles south of Salzburg below Hitler's residence at Berchtesgaden. When the village was overrun in April 1945 by Moroccan troops, their commander, Commandant Roger Dubaquier, who was Le Chene's brother-in-law, assembled von Neurath's decorations and later gave them to the Moroccan Goum Museum, except for the Grand Cross, which he gave to Le Chene as a mark of the respect that France had for him.

The Hubble Chess Set

Captain Desmond Hubble transferred from the Royal Artillery into the Intelligence Corps in December 1939 and served with No. 19 British Military Mission (West Africa) between 1941 and 1943. Joining the SOE French Section, during the night of D-Day he parachuted into the Ardennes region as part of Inter-Allied Mission 'Citronelle', but was captured six days later. After being transferred to St Quentin Prison in Paris, in August, he was one of thirty-seven SOE agents transferred to Buchenwald Concentration Camp, where Hubble's Battle Dress, with his Intelligence Corps insignia, and his chess set inspired the prisoners. In early September, thirty-one of the prisoners, including Hubble, Captain Frank Pickersgill (Canadian Intelligence Corps) and the Canadian Captain John Macalister (Intelligence Corps), were executed hung from hooks. Shortly before the end of the war, Wing-Commander Forest Yeo-Thomas, also French Section, escaped from Buchenwald but was recaptured and sent to a prison camp of Frenchmen. He was able to prove he was not a stool pigeon by showing the chess set with its inscription 'Made in England'.

Medals of Major Alistair MacDonald MC Intelligence Corps

In July 1944, Major MacDonald co-ordinated French Resistance operations in Central France. In October, in command of the 'Cherokee' Mission, he parachuted close to the town of Biella, north-west of Milan in Piedmont, Italy. Among the 'Banda Biella' partisan group were three prisoners who had ignored the 1943 Stand Fast order to Allied prisoners when Italy surrendered. In early 1944, MacDonald organised several supply drops that enabled the 'Banda Biella' to harass the retreating German Tenth Army lines of communication to Germany until he was captured in a German counter-intelligence sweep. When he escaped from Gestapo HQ and reached Switzerland, he was replaced by Lieutenant Amoore, who had previously been an intelligence officer with the Polish Corps at the Battle of Monte Cassino in 1944. As the Germans fought to protect their lines of communication with anti-partisan operations, Amoore was forced into hiding for several weeks. He then disrupted German logistics by destroying two railway bridges and forcing the enemy to transfer supplies from the railway to trucks and then back onto trains. Daily supply drops enabled the partisans to close the railway through the Aosta

Valley and ambush the convoys on the roads. On 2 May, the day that the German Army in Italy surrendered, Captain Amoore accepted the surrender of LVII Corps. His medals include the MC, Cross of Valour (Poland) and Military Cross for Valour (Italy).

MacDonald's Liberation Medal, front

MacDonald's Liberation Medal, reverse

Medals of Captain John Amoore MC Intelligence Corps

Dead Letter Box, in the closed position

A 'dead letter box' (DLB) is a place where agents and contacts can leave and collect messages without the sender and recipient meeting. The dog DLB was found during a FS counter-intelligence operation in Allied-occupied Italy and contained microfilm ready to be collected by a German agent.

Dead Letter Box, open.

OP-3/Type 30/1 Wireless and Power Pack

The OP-3/Type 30/1 was a clandestine wireless receiver conceived by the Polish Military Wireless Unit, then based in Stanmore. Consisting of the receiver and power supply, it was designed to be small enough to be attached to a belt under a coat. 105 sets were dropped to occupied Poland and a few in 1944 to occupied Denmark.

Medals of Sergeant Leslie Gilbert MM

Sergeant Gilbert was one of several German and Eastern European refugees serving in the Auxiliary Military Pioneer Corps who were transferred into a special Intelligence Corps unit in 1944 specifically to infiltrate enemy lines and collect intelligence. While supporting the US 84 Infantry Division near Erkelemz in February 1945, although pinned down by heavy enemy fire, he crawled near an 88 mm gun battery and called down fire onto his position. Three weeks before Operation *Plunder* (the crossing of the River Rhine), during the night of 28 February, he and Sergeant Harry Saunders crossed the river posing as anti-tank gunners sent to collect spares from an Essen factory. They spent the next week in the city but as they were returning to the Rhine, a German military police sergeant-major, suspicious of their documents and lack of spare parts, locked them in a guardroom. Next day, they convinced a military police major they had lost the spares when the boat taking them across the Rhine was hit by the wreckage of an aircraft and were released. They were debriefed at HQ Twenty-First Army Group. Both NCOs were awarded Military Medals for their gallantry.

Medals of Colonel John (later Sir John) Cecil Masterman OBE

John Masterman was interned in Germany during the First World War. Commissioned into the Intelligence Corps during the Second World War and seconded to MI5, he chaired the Double-Cross Committee that turned captured agents to feed disinformation to their German controllers. By 1941, almost all German agents in UK were under MI5 control. During the planning for the Allied landings in Normandy, in Operation *Fortitude*, turned agents were used to deceive German commanders into believing that a phoney army in Eastern England would land in the Pas de Calais, a deception so effective that the German High Command delayed sending fifteen reserve divisions to Normandy. When the Germans launched V-bombs against London in 1944, Masterman used the agents to deceive the Germans into aiming the missiles away from the capital.

Air Photograph of Rouen, 1943

From the interpretation of the titling strips and the 542 (Photographic Reconnaissance) Squadron log, the photographs were taken on a sortie flown by Flying Officer Alec Paus over Dieppe, Rouen and Billancourt between 16.35 and 19.05 on 4 April 1943 after a bombing raid. The focal length of the camera was F135. 542 Squadron was formed at RAF Benson with two flights of Spitfire IVs on 19 October 1942. Also found with the air photographs was a wartime black and white map prepared by 14 Regiment RE.

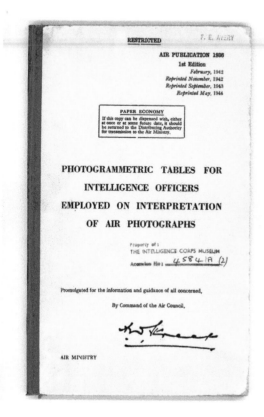

Photogrammetric Tables for Intelligence Officers on Interpretation of Air Photographs

First published by the Air Ministry in February 1942, the tables provide mathematical calculations in conversions between imperial and metric units, taking into account European and Greenwich meridians. The tables would have been used during the Anglo-American Operation *Bodyline*, later *Crossbow*, against the German V-Weapon programme. The tables were superseded by slide rules that enabled the Photographic Interpreters to calculate scales and then set results on another scale, from which the image measurement could be set to produce the ground measurement.

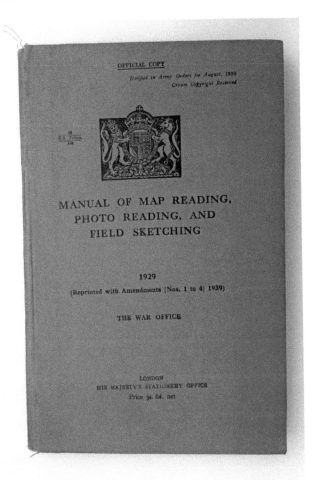

Manual of Map Reading, Photo Reading, and Field Sketching (1939)

Updated from the iconic 1929 publication and specific for the Army and RAF, the manual provided instructions for Field Sketching, the importance of which could 'not be emphasised', map reading and Air Photographic Reading. Place names for regions that did not use the English alphabet – for instance, Japan – were to be spelt phonetically on principles used by the Royal Geographical Society.

German Intelligence Medal

The medal and its envelope were found in a German Army intelligence office in Lille in July 1944. On the reverse, it is inscribed '*Fur Arbeit Zum Schutze Deutsch Lands*' (Work to Protect German Lands).

Ghent Liberation Medal

The medal was presented to Captain H. Russell-Ross for his part in the liberation of the city in September 1944.

After transferring to the Intelligence Corps from the Royal Artillery, Captain Henry Russell-Ross was the HQ VIII Corps Counter-Intelligence Staff Officer in 1944. Issued with an arrest list of enemy suspects in the Belgian city of Ghent, he joined 'Ghent Force' which had been assembled from the 7th Armoured Division and tasked to liberate the city, contacted the Belgian Resistance and organised arrests. However the city council thought that he was the Town Major and appointed him Mayor. He continued making arrests until the Second Army Field Security Reserve Detachments arrived. Russell-Ross was involved in the arrest of several senior German Army and Navy officers after the German surrender in 1945.

Goering Envelope Stand

The plaque reads, 'Liberated from the Berlin Office of Hermann Goering by an officer of the Intelligence Division of the Allied Control Commission'.

Inkwell

The inkwell was found in the *Geheime Feldpolizie* HQ in Berlin in 1945. A swastika has been scratched on the left-hand lid. The GFP had responsibility for military counter-intelligence, protective security and counter-resistance, and was an executive branch of German military intelligence. Its personnel regularly wore civilian clothes and were trained in source-handling, surveillance and interrogation.

The Belsen Typewriter

Lieutenant Derrick Sington (Intelligence Corps) served with 14 Amplifier Unit and was one of the first to enter Belsen Concentration Camp in April 1945. Amplifier units were involved in Psychological Warfare Executive broadcasts. When he was instructed to compile lists of German prisoners and inmates, the only typewriter available was a German one that included the 'SS' when the number 2 key was tapped plus SHIFT. Inside the typewriter cover is official authority (see overleaf) for Lieutenant Sington to use it in case he was accused of theft.

Sington Permission

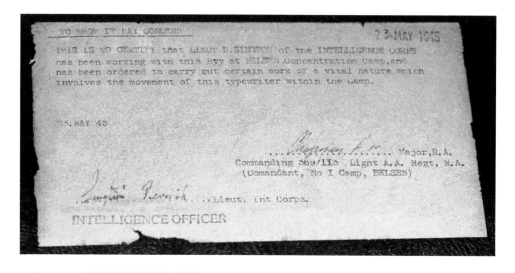

TO WHOM IT MAY CONCERN   2 MAY 1945

THIS IS TO CERTIFY that LIEUT D.SINGTON of the INTELLIGENCE CORPS
has been working with this Bty at BELSEN Concentration Camp, and
has been ordered to carry out certain work of a vital nature which
involves the movement of this typewriter within the Camp.

25.MAY 45

                                        .... Major, R.A.
                            Commanding 569/115 Light A.A. Regt, R.A.
                            (Comandant, No I Camp, BELSEN)

.... Lieut, Int Corps.

INTELLIGENCE OFFICER

Reich Labour Service Jacket

From June 1935, men aged between eighteen and
twenty-five years were expected to serve six months
in the *Reichsarbeitsdienst* (Reich Labour Service;
RAD) before enlistment into the Armed Forces.
The organisation was established to mitigate
the impact of unemployment, to militarise the
workforce, to provide auxiliaries and to enable
Nazi indoctrination. During the Second World War,
women were conscripted into the RAD.

Set of Four German Swords

The top dagger is for a
Luftwaffe officer. The smallest,
complete with a swastika
inscribed on the hilt, was found
on a prisoner of war in a prison
camp search. The remaining
two are unidentified.

General Eicke Sword

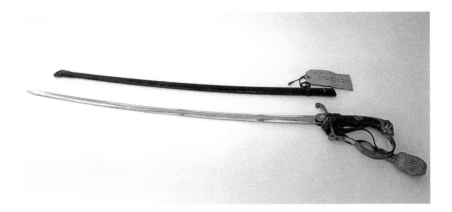

Hilt of the Eicke Sword

The Nazi Honour Sword owned by SS General Theodor Eicke was found, minus its scabbard, at his home near Sachsenhausen Concentration Camp by Dr Hans Hers, a Dutch intelligence officer, who presented it to the Museum in 1995. Eicke established the concentration camp system and then commanded the SS Totenkopf Division, which was one of the most effective German formations on the Eastern Front, but noted for its brutality. One unit had murdered ninety-seven British prisoners at Le Paradis in 1940. Eicke was killed in Russia in 1943. The hilt clasp of oak leaves indicates the rank of General and is formed by a downward-facing snake. The black and silver colours are symbolic of the SS. The emerald green eyes of the lion were symbolic of the Gestapo and Nazi State Security.

Political Warfare Executive

The British formed the PWE in 1941 to produce and disseminate both 'white' (true) and 'black' (inaccurate) propaganda in order to damage enemy morale and sustain civil morale in occupied countries.

Forged Ration Coupons

The pamphlet 'signed' by Adolf Hitler claims that Germany had suffered 3,608,200 soldiers killed and 9,114,400 wounded between 1939 and 1943.

Forgery of Casualty Statistics

The forged coupons are 'a gift' from Hitler and were designed to undermine the rationing of food. Both examples were printed by Leagrave Press and dropped into Germany.

Occupation, Cold War and British Counter-Insurgency Operations and Wars

Military Civilian Government Service Dress Jacket

Military Civilian Government Blue Battle Dress Jacket

The jackets were worn by an Intelligence Corps officer attached to the Military-Civilian Government (MCG) in Occupied Germany. Its role was to govern Germany, rebuild social and industrial infrastructure, arrest war criminals and resettle the millions of displaced people. One war criminal arrested by 45 FSS and 1004 FSRD in late May 1945 was Heinrich Himmler, posing as 'Sergeant Hitzinger'. Since the MCG was a civilian organisation, all military insignia were removed, including replacing brass buttons with leather ones. The Intelligence Corps provided a security organisation for the MCG. The Control Commission Germany badge on the left sleeve was adapted from GHQ Troops, 21st Army Group. It eventually took over from the MCG. The blue Battle Dress tunic was worn when there was a probability of clothes becoming dirty; otherwise khaki Battle Dress was worn.

Handcuffs

The handcuffs are inscribed with the names of three Germans arrested by 92 FSS for war crimes. On 4 March 1946, SS Lieutenant-Colonel Rudolf Hoess, the Auschwitz concentration camp commandant, was arrested while posing as a discharged sailor on a farm near Flensburg and was later hanged outside his office in Auschwitz. Arrested on 31 March 1946 was SS Captain Hans Bothman, the last Chelmno extermination camp commandant and commander of the SS Special Detachment responsible for exterminating Jews in Lodz. He committed suicide. SS Captain Rudolf Renner was arrested on 27 June 1946 and extradited to Denmark where he served eight years of a twenty-year sentence.

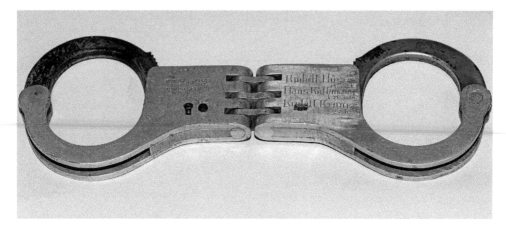

Medals of Colonel Alexander Paterson Scotland OBE

The son of David Scotland, Colonel Scotland developed the Prisoner of War Interrogation Section (PWIS) network of Command interrogation centres in Great Britain, the most well-known of which was the London Cage in South Kensington. In 1945, the PWIS accommodated the War Crimes Investigation Unit (WCIU) and among the accused were Nazi officers and officials, including police thought to have been involved in the murder of fifty Allied airman after the Great Escape. Several prisoners claimed London Cage breached the Geneva Conventions, including using violence, a claim rejected by Scotland as counter-productive, 'particularly with some of the toughest creatures of the Hitler regime'. Scotland refused the Red Cross access to London Cage on the ground that the prisoners were neither civilians nor criminals within armed services. Scotland would later write a controversial book about the London Cage. Other medals are the British War Medal, Victory Medal with MID Oakleaf, 1939–45 Star, Defence Medal, Victory Medal and US Bronze Star.

'Smokey Joe', Vienna

In 1945, Austria was divided between the four Allied powers. When MI6 discovered that Soviet military telephone lines passed through the British sector between 1948 and 1951 in Operation *Conflict*, 291 FSS manned the listening post, which became known as 'Smokey Joe's' due to the thick fog of Woodbines and Players cigarettes. The model depicts (top left) the entrance to the listening post protected by a guard armed with a Sten gun; (lower right) rest area; and (lower left) the 'telephone tap'. The intercept was lifted when Soviet forces left Vienna. Operation *Conflict* became the model for a joint British/US operation in Berlin also using a tunnel, but it was compromised in 1954 by the spy George Blake.

Post-1945 Home Guard

With the increased threat from the Soviet Union, in 1951 the Home Guard was reformed with an expected initial establishment of 170,000 men; however, only about 23,000 had joined, with a further 20,000 on a Reserve roll. The Battle Dress jacket was worn by a wartime and post-war Intelligence Corps lieutenant serving in the Home Guard. The force was disbanded in 1957, only to reform as the Home Service Force (1982–92), which had a specific role to protect military key points from enemy special forces. At least two former Intelligence Corps members enlisted.

Medals of Staff Sergeant Clifford Jackson DCM MM

During the Second World War, Jackson was awarded the DCM and Silver Star (US) for gallantry in Italy. Transferring to the Intelligence Corps, between October 1951 and May 1953 he commanded the Special Detachment, 1st Commonwealth Division, recruiting, training and handling agents operating behind North Korean lines. While the nature of the operations remain classified, Jackson's commitment enabled the Division to benefit from the intelligence he supplied. He was awarded the MM.

Medals of Major-General F. H. M. Davidson CB DSO MC

As Intelligence Corps Colonel Commandant (1952–60), Davidson oversaw the transfer of the First 100 and won official acceptance that the Corps had been formed on 19 July 1940. Commissioned into the Royal Artillery in 1911, he was wounded four times and awarded the MC and Bar during the First World War. During the Second World War, while Director of Military Intelligence (1940–44), he realised the value of a permanent Intelligence Corps. His call, 'Onwards and Upwards!', resonates today.

89 FS Para Battle Dress Jacket, Worn by a Lance-Corporal

The Intelligence Corps' strong link with the Airborne Forces extends to June 1942, when 89 (Airborne) FSS was formed. The Section fought at Arnhem as infantry. One NCO who escaped from the siege was found dead from exhaustion in his bed.

**Captain Isaacs escorting
Japanese *Kempei Tai*
to a prison camp,
Singapore, 1945**

Medals of Lieutenant-Colonel Richard Isaacs MBE MC Intelligence Corps

One of the First 100, Richard Isaacs was appointed the first Corps Lieutenant-Colonel. He had enlisted into the Grenadier Guards before the war and, on transferring to the FSP, was evacuated from Dunkirk. Commissioned in 1941, he commanded 565 (5th Division) FSS in Burma from 1944 to 1945 and was the only Field Security Officer to be awarded the MC. When Singapore was liberated, he escorted the *Kempei Tai* to a prison camp and was presented with the Japanese sword by 5th Division. Thereafter he served in the Malayan Emergency and West Germany, and became involved in Psychological Warfare.

Japanese Sword

Intelligence Corps Sweater

When in the late 1960s the Army Dress Committee permitted Corps and Regiments to wear distinctive sweaters, the Intelligence Corps selected bottle green as its colour. When the Committee rescinded the decision in 1972, the Corps was one of those that reverted to the issue Heavy Duty sweater.

No. 1 (Formal) Dress – Blues

The mannequin shows an Other Rank wearing the distinctive Intelligence Corps Green Beret. On the upper arm of the tunic is the Sykes dagger insignia worn by soldiers who passed the commando course. The Corps association with Commando Forces stretches back to 1942, when several officers were involved in the raising and training of No. 10 (Inter-Allied) Commando. In 1945, its 3 (British) Troop, which consisted largely of East European refugees serving under anglicised names, were transferred as a reinforcement to Intelligence Corps operations.

No. 2 Dress (Temperate Parade)

The mannequin shows an Intelligence Corps Warrant Officer Class 1 wearing No. 2 Dress. This macramé green lanyard worn on the left shoulder is awarded to soldiers accepted into the Intelligence Corps.

No. 10 (Mess) Dress Jacket

The green represents the traditional colour of military intelligence and the grey lapels commemorate the connection of the green and grey brassards won by the First World War Intelligence Police and its successors.

Notes of Interest (1944)

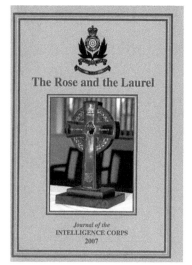

The Rose and Laurel (2007)

During the Second World War, the Intelligence Corps produced a short newsletter entitled *Notes of Interest* that summarised the activities of FS units. The 1944 edition lists those known to be killed in action, wounded and prisoners of war, including in 'Japanese hands'. By 1964, *The Rose and Laurel* emerged as the Corps Journal. The 2007 edition front cover shows the Celtic cross commemorating all Intelligence Corps killed during Operation *Banner* in Northen Ireland, including Corporal Paul Harman, who was ambushed in Belfast while driving a civilian vehicle in December 1977, and those killed when a RAF Chinook crashed at the Mull of Kintyre on 2 June 1994.

The Development of Military Intelligence

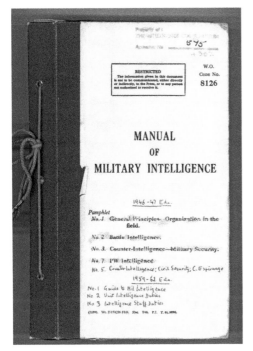
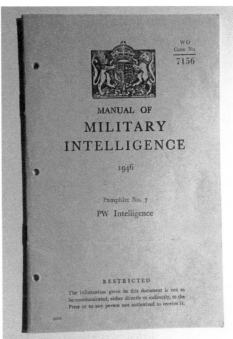

Above left: *Manual of Military Intelligence* (1946/47 & 1959/62)

Above right: *Pamphlet No. 7 (PW Intelligence)*

Below: Command Post diorama

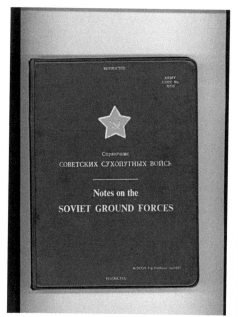

Above left: Notes on the Soviet Ground Forces

Above right: Notes on British Army Organisations and Equipment

The *Manual of Military Intelligence* 1946/47 gives guidance for Staff and unit Intelligence officers by covering General Principles of Intelligence in the Field and describing intelligence support to tactical air forces and naval task forces in the Pacific Islands and jungle, and covers the role of Signals Intelligence and Imagery Intelligence.

Pamphlet No. 7 (PW Intelligence) gives principles for dealing with the interrogation of captured enemy personnel, civilians and the debriefing of British prisoners, escapers and evaders. It highlights the necessity for resistance-to-interrogation training. The pamphlet gives instruction from selecting prisoners for battlefield interrogation through to the high-level Combined Services Detailed Interrogation Centres. It concludes by listing several articles of the 1929 International Convention Relative to the Treatment of Prisoners of War and lists the signatories to include Germany and Italy but not Japan.

The 1959–62 manual expanded operations to include Counter-Insurgency intelligence lessons learnt from the Malayan (1948–60) and Kenya (1952–56) Emergencies.

The diorama portrays an intelligence cell in West Germany during the 1970s/1980s. 1:50,000 maps produced by the Royal Engineers show enemy forces marked in red; friendly forces in blue; obstacles, e.g., minefields, in green; and NBC in yellow. Included on the table is an orange map marking template, a ruler inscribed with Soviet artillery ranges, *Notes on the Soviet Ground Forces* in the green cover and a S-10 respirator for quick use. Intelligence Corps were also trained in clerical skills and thus the typewriter.

Notes on the Soviet Ground Forces was the 'intelligence bible' and described not only conventional tactics, but also specialised warfare such as air assault; amphibious and river crossing; winter warfare; fighting in built-up areas, forest, mountains and desert; and the life and political culture of the soldier. Also detailed were organisational charts, command and control, and equipment – *Tactics of the Soviet Ground Forces* gave more detail. Intelligence courses also taught British Army organisation and equipment.

The mannequin wears a British NBC suit designed to give protection from radioactive, biological or chemical substances for extended periods, which could be donned quickly over a uniform. The British knew it as a 'Noddy suit' because of the pointed hood worn by the fictional character. The term 'NBC' has been replaced by 'CBRN' (chemical, biological, radiological, nuclear).

Air Army Photographic Interpretation Kit

While 6 Intelligence Company provided Imagery Intelligence for BAOR, Intelligence Sections at all levels also carried out photographic interpretation. Designed to specifications set by Major L. Skelton (Royal Artillery), an Army liaison officer at JARIC designed equipment carried in a robust attaché case that included a stereoscope, slide rule and arithmetic tables.

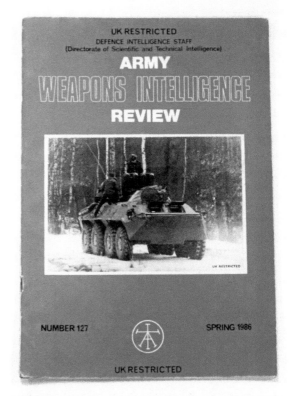

Army Intelligence Weapons Review

The Ministry of Defence regularly published Army Intelligence Reviews covering a wide range of subjects. Commercial publications relating to maritime, ground and air equipment, such as the Jane's publications, were also valued.

THREAT

THREAT was a six-monthly periodical produced by 7 Intelligence Company to educate the BAOR on the military threat from the Group of Soviet Forces in East Germany. The exhibit shows a handy pocket-sized *aide memoire* showing Soviet organisations, uniforms and equipment.

When the British sent troops to liberate Kuwait, 7 Intelligence Company produced *THE RAT* (an anagram of *THREAT*) to describe Iraqi forces

THE RAT (Break-Up of Yugoslavia Special) produced by the Command Intelligence Training Unit in Bulford. Also exhibited is *Defence Recognition* (former Yugoslavia Air and Ground forces) and the standard Pink Card giving guidance on opening fire and the use of lethal force.

BRIXMIS

Diorama

The diorama shows two BRIXMIS observing Soviet activity. In September 1946, the British and Soviet commanders in Germany agreed under the Robertson-Malinin Agreement to exchange military liaison missions to foster good military relationships, repatriate prisoners of war, displaced persons and deserters, search for and extradite war criminals, register war graves, and settle border disputes. The Missions were respectively known as BRIXMIS and SOXMIS. While similar arrangements were agreed between the Soviet, French and US commanders, BRIXMIS had almost as many liaison staff in the Soviet Zone as the other two missions combined. The agreements ended on 2 October 1990.

G-Wagon

BRIXMIS used several types of vehicle as Tourers, including Opel saloons. In the mid-1980s, the Mercedes Gelaendewagen, known as the 'G-Wagon', replaced Range Rovers. With an excellent cross-country capability, its large windows sometimes attracted aggressive attention from Soviet and East German military. One Opel was rammed by a lorry with such force that its Intelligence Corps driver spent several days in an East German hospital with serious injuries.

The ZIL-131

When Germany was re-united, the ZIL-131 radio truck was given to the British by the East Germans, complete with documentation. The general purpose, 3.5 ton, 6 x 6 truck was designed in the Soviet Union.

JARIC Recognition Training Models

T-55 Tank Mine-Rollers

T-55 main battle tanks that first appeared in 1958 had been modified from the T-54 series, with NBC protection and a brand new engine. T-54s modernised to the T-55 standard were known as T-54/55s.

PMP Bridging

The PMP Floating Ribbon Bridge was designed by the Soviet Army during the Second World War. With a payload of 60 tons, lorries parked alongside the riverbank, the pontoons roll off and, unfolding automatically, were towed into position by motorboats. Thirty-two pontoons can bridge 390 m of water. Four are shore pontoons.

Soviet Airborne Forces Jacket

A lightweight summer jacket worn by Soviet Airborne Forces (*Vozdushno-desantnye voyska*), identifiable by the light blue of the airborne Arm of Service collar gorget and rank shoulder boards. The sleeve badge shows a Hammer and Sickle inside a red star and a parachute flanked by two transport aircraft enclosed in a blue and yellow shield.

Soviet Arm of Service Badges

Used for a training film made in the mid-1980s, the 'spoof' and real Soviet Arm of Service are (left to right, top to bottom): 'spoof', Tank, Air Force, Artillery, Motor Transport, Chemical Warfare, Military Engineers, 'spoof', Motor Rifle and Border Force. The badges are ORs' cap badge and officers' cap badge. In the centre is a badge designed to represent the KGB. To its left is an officer's cap badge and an 'upside down' Polish airborne badge.

Soviet Submariner's Greatcoat

The woollen naval greatcoat carries a black shoulder board with the letter 'K' indicating a cadet. The badge signifies submarine service.

Chinese Border Force Jacket

The Chinese Border Force is part of the Chinese Armed Police and manned posts on the border with Hong Kong. The shoulder insignia shows the Chinese Community Party red star surrounded by four smaller stars, representing the peasantry, working class and urban and national bourgeois.

The Development of Military Security

Pamphlet 3 (Military Security), Manual of Military Intelligence 1946/47, superseded the 1943 issue and covered Field Security in the advance and the withdrawal, and also in combined operations, counter-intelligence, civil security and counter-espionage, censorship, signals security, Travel Control Security and 'precautions to be taken by British troops in the event of capture'.

Military Security Instructions (MSI) set the minimum standard that all Army units and civilian components were expected to achieve. Every organisation had a security officer trained by the Intelligence Corps answerable to the most senior Army officer and civilian. They were expected to compile comprehensive local security instructions based on MSIs and conduct security education and training. Intelligence Corps security units in every Command inspected the state of security, with rolling programmes of checking classified document control, weapon and equipment security and access control. Inspection reports were formal and listed conclusions and recommendations, and were circulated to the General Staff. The sections also conducted counter-intelligence investigations and gave architect liaison advice new builds and refurbishments. The growth in national and international terrorism in the 1970s and the appearance of Information Technology led to MSIs being replaced by the *Manual of Army Security* (MAS). The strategy led to military security being embedded throughout the Army, to the extent that it was largely prepared when the IRA opened their campaign in the early 1970s and did not face similar breaches to those experienced by other government departments and the security industry.

Below left: *Military Security Instructions* (1956)

Below right: *Military Security Instructions* Issued to the Counter-Intelligence Platoon, Hong Kong

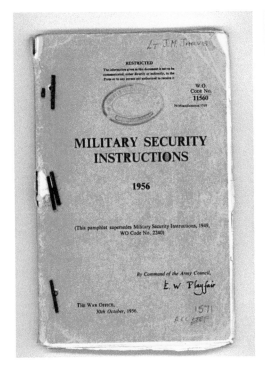

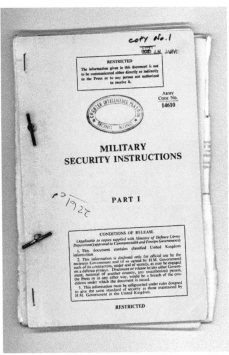

Below left: *War Office Military Security Instructions* (Red Cover)

Below right: *BAOR Security Instructions*

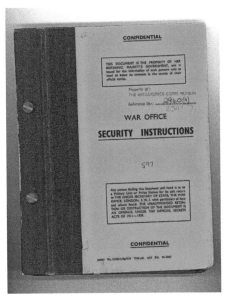

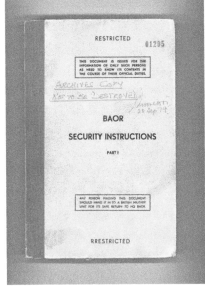

Manual of Army Security (1977)

Poster, *Our Security Depends On You*

The HMSO booklet is one of two with the same title that describes the subversive and espionage threat from the Russian Intelligence Services of the GRU (military intelligence) and the KGB (state intelligence). Case histories describe the motivations for espionage, which include ideology, mercenary reasons and social pressure. Vetting limits access.

Booklet, *Their Trade is Treachery*

Both booklets describe methods used by Communist intelligence services to recruit spies and gives advice on how to avoid being recruited. *Treachery Is Their Trade*, of fifty-nine pages, was distributed in 1964 after several embarrassing defections and at a time when the Government feared the Soviet spy machine. Published in 1975, *Treachery Is Still Their Trade* extracted more lessons for the Government and Armed Forces by describing fifteen cases and highlighting the threat against scientific, technological and industrial information. The booklets were given to Intelligence Corps as part of protective security training.

Personal Interview Kit

All Intelligence Corps were trained in security interview techniques. This led to some assembling personal interview systems. The kit includes a cassette tape recorder and cassettes, listening device, cable to enable covert recording, torch, utility tool and notebook.

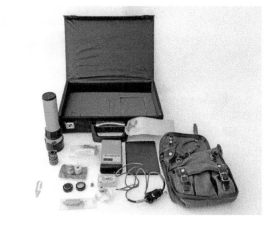

Flat Iron

The flat iron was used by the British Security Service Organisation (BSSO) in Berlin to reseal intercepted letters. The BSSO had strategic responsibility for protecting the British Army of the Rhine from Eastern Bloc infiltration and domestic and international terrorism.

Medals of Lieutenant-Colonel Colin Barnes OBE

The Intelligence Corps seconded officers and NCOs to several foreign armies, where they were expected to integrate with their hosts by wearing the national uniform. Colin Barnes transferred from the Royal Artillery into the Intelligence Corps in 1945 and spent most of his career seconded to Middle East armies. He became an influential adviser to several regional heads of state. The medal array also shows a 1939–45 War Medal, GSM with Palestine, Arabian Peninsula, Ethiopia clasps, Star of Ethiopia, Abu Dhabi DSM, Oman DSM and Sultan of Oman's Commendation and Bar, Oman GSM with Dhorar clasps and Oman's Victory and Peacekeeping Medals. Barnes was a Friend of the Royal Geographical Society.

Royal Brunei and Malay Regiment Mess Jacket

The jacket was worn by an Intelligence Corps NCO seconded to the Royal Brunei and Malay Regiment for two years.

Medals of Sergeant Les Maisey (1913–78)

Les Maisey joined the Army aged sixteen and, while serving with the Gloucester Regiment, took part in the epic 1,000-mile retreat from Rangoon in Burma to India in 1942. Transferred to the East Yorkshire Regiment, he served in Palestine, Egypt and Austria after the war. Self-taught in German, he transferred to the Intelligence Corps in 1953, was posted to Singapore and closed 904 (Korea) FSS in 1957. Discharged in 1965, he was employed by the Intelligence Centre in Ashford in the Quartermasters' and Sergeants' Mess and was frequently called on for character parts in instructional wings. He was awarded the Meritorious Service Medal, the GSM and LSGC and bar representing thirty-seven years' uniformed service.

Medals of Major Robin Rencher BEM

After emigrating to Australia in the mid-1960s, Robin Rencher enlisted into the 6th Royal Australian Regiment and, while serving with D Company, was wounded during the Battle of Long Tan in South Vietnam in August 1966. D Company was awarded a US Presidential Unit Citation. After transferring to the Australian Intelligence Corps, he was wounded a second time in Vietnam. He then transferred to the Intelligence Corps in 1971 and served in Northern Ireland, the UK, West Germany and in the Falklands Campaign.

WO2 Rencher being awarded the LSGC Medal by General Gow

Sergeant Rencher being awarded the LSGC (Army) by General J. M. Gow, Colonel Commandant Intelligence Corps 1972–86. On his sleeve is the US Presidential Unit Citation.

Cyprus

EOKA Pamphlet

In 1955, *Ethniki Organosis Kyprion Agoniston* (EOKA; National Organisation of Cypriot Fighters) conducted a four-year insurgency campaign in Cyprus that led to the island dividing into two territories. Militants then formed EOKA-B, which periodically surfaced with terrorism against British Forces, Cyprus. The pamphlet was found in Limassol in 1986 and calls for the removal of the two British Sovereign Base Areas during a period when Palestinians attacked a British base with mortars and AK-47s.

Operation *Banner* – Northern Ireland

Action Cards

Operation *Banner* was one of the counter-insurgency campaigns in which Service personnel were issued with instructions and guidelines on dealing with the incidents. These included the Yellow Card *(Instructions by the Director of Operations for Opening Fire in Northern Ireland)*, the Blue Card, which gave instructions for making arrests, and the Buff Card, which dealt with terrorist ambushes, bombs and booby traps.

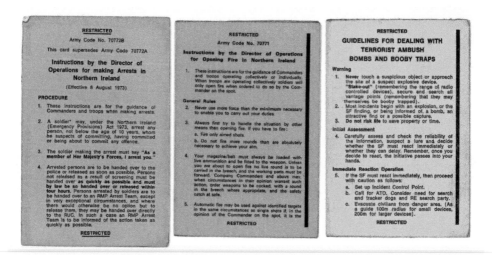

Terrorist Weapons

Flanked by a *Quick Guide to Terrorist Weapons,* produced by Weapons Intelligence, are two single-shot, short-range 'zip' pistols. Such weapons were usually manufactured in garages and workshops.

Soviet 7.62 mm Kalashnikov AK-47 Assault Rifle with Folding Butt

Many were smuggled to Northern Ireland by ship from Libya. The AK had an automatic capability and a distinctive curved thirty-round magazine. The Museum also has an example of an IRA 'Barrack Buster' mortar of a large gas canister fired from a tube against Security Forces bases.

Walther 5.56 mm PKK

In several British counter-insurgency campaigns, members of the Security Forces wearing civilian clothes either on duty or recreation frequently carried personal weapons. In Northern Ireland, this generally consisted of either the Walther 7.65 mm PPK made famous by James Bond, or M1935 9 mm High-Power Browning. Both pistols had an effective range of 30 m and, while the PPK magazine holds seven rounds, the capacity of the Browning is thirteen rounds.

Republican and loyalists in prisons frequently regarded themselves as prisoners of war and, as part of a welfare programme, friends, families and prisoners sold inscribed handkerchiefs, leather holsters and wristlets, and paintings.

IRA Handkerchief Roll of Honour

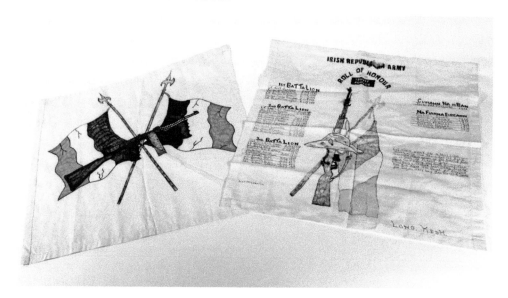

Loughgall T-Shirt

The item commemorates the deaths on 8 May 1987 of eight of the IRA who were ambushed as they planted a bomb at the Royal Ulster Constabulary Police Station, Loughall. Two civilians were also killed.

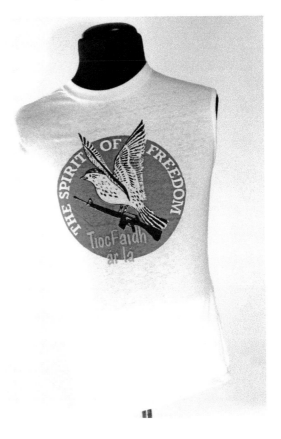

Sinn Fein T-Shirt

The t-shirt depicts a dove (representing peace) and an Armalite. After the electoral success of Bobby Sands in 1981, Danny Morrison of Sinn Féin asked, 'Who here really believes we can win the war through the ballot box? But will anyone here object if, with a ballot paper in this hand and an Armalite in the other, we take power in Ireland?' Such handkerchiefs had Documentary Intelligence value for identifying members of paramilitary organisations.

UVF Compound Nominal Roll

Operation *Corporate* (Falklands War)

Armada Argentina *Reglamento del Grupo de Infanteria de Marina en el Combate*

The Handbook was captured in South Georgia in April 1982 and was of Document Intelligence value because, for the first time, the Landing Group in the Task Force had current intelligence on the Argentine Armed Forces. The manual includes the principles of Argentine marine tactical and organisational philosophy. Chapter 3 relates to field security, the treatment of prisoners of war and intelligence operations against guerrilla forces.

Argentine Sheath Knives

An Argentine corporal being repatriated to Argentina gave the top knife and sheath, presented to him on a promotion by his colleagues, to an Intelligence Corps NCO serving with HQ 3rd Commando Brigade. The lower knife and sheath was found on the body of an Argentine commando sergeant on 29 May 1982 on Mount Kent after he attacked a position occupied by an Intelligence Corps lance-corporal serving with the SAS. The knuckle-duster handle has been damaged by a bullet. The lance-corporal was wounded in the battle and was evacuated ten hours later under cover of darkness.

Collection of Argentine Equipment

The items include a inscribed white naval towel, marine infantry helmet, pair of goggles frequently used by Argentine servicemen, set of mess tins on top of a leather pouch, badges of rank, unit shoulder flash, and the helmet and holster worn by the pilot shot down during the Battle of Goose Green on 28 May 1982.

Coalition Operations

Operations *Granby* (Liberation of Kuwait) and *Telic* (Iraq)

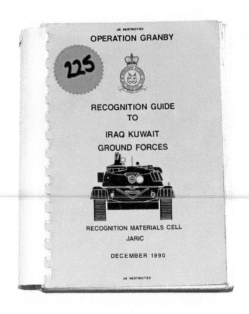

Recognition Guide to Iraq/Kuwait Ground Forces

Produced by the JARIC Recognition Materials Cell in December 1990 for use by the British forces in Operation *Granby*, it has photographs and sketches of equipment used by the Iraqi Armed Forces, which included some Western equipment.

Iraqi Map Case and Marked Map Trace

Document Intelligence found by British troops in an Iraqi HQ.

British Clansman VHF 638 Radio

A radio sold to Iraq and captured by British soldiers. With it was a card that indicated the Iraqi concern about Coalition Signals Intelligence by using codes to describe events: e.g., Book = *Enemy tanks near Divisional HQ*; Dates = *Our position penetrated*; Door = *Enemy surrounding us.*

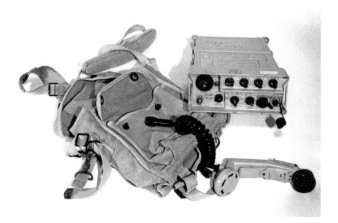

Iraqi Military Police Steel Helmet in the Soviet style with a red band and a star on a white background. Also the standard Iraqi helmet

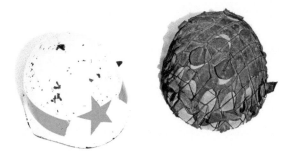

Soviet 40 mm Rocket-Propelled Grenade-7 (RPG-7), captured in Iraq

Weighing 7 kg, the portable, shoulder-launched, anti-tank RPG-7 has an effective range of 200 m. Rugged and cheap, it and its predecessor, the RPG-2, are widely used by guerrilla forces and terrorist movements, including those in Northern Ireland.

T-59 Tank

The Chinese T-59 medium battle tank captured in Iraq. Weighing 36 tons and crewed by four (commander, gunner, radio operator/loader and driver), its main armament is a 100 mm gun with secondary 7.62 mm turret machine gun and cupola 7.62 mm machine gun for air defence.

Bust of Saddam Hussein

Combat Indicator Guide (2000)

Military forces frequently leave a signature of their intentions – for instance, the assembly of bridging equipment means a river crossing. This is known as a combat indicator. The handbook issued by the Army Combat Intelligence Branch is designed to give ISTAR (information, surveillance, target acquisition and reconnaissance) assets the ability to interpret enemy intentions, capabilities and vulnerabilities, and to help identify concealed diversions.

Operation *Grapple* – The Balkans

Operation *Grapple* was a British support to UN and NATO peacekeeping forces as Yugoslavia split into Serbia, Croatia and Kosova while the authority of the Soviet Union melted.

The opponents based their military structures on the former Yugoslavian Army and the only way they, and indeed peacekeepers, could identify the factions was by large badges. In the centre is a Yugoslavian Army cap.

Operation *Herrick* – Afghanistan

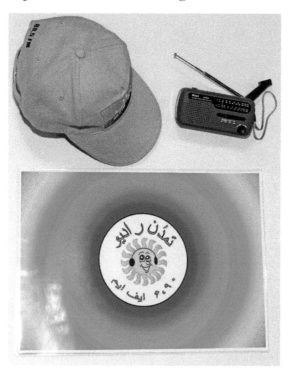

Psychological Operations Transistor

Psychological Operations is an intelligence asset designed to undermine enemy subversion and influence the values and motives of target audience. Historically, leaflets were dropped from aircraft, and aircraft equipped with loudspeakers were used. The widespread use of Information Technology and social media has allowed techniques to evolve. For communities that lack electricity, small self-winding transistor radios allow the authorities to keep communities 'on side'. On the cap badge is the Psychological Operations frequency.

Psychological Operations (white)

The education of children is important, and toys and puzzles helps broaden their outlook. The exhibit is a jigsaw of the provinces of Afghanistan, which, when completed and reversed, shows the national flag of the country. Posters warn against picking up explosive devices.

Company Operational Intelligence Support Handbook

The handbook advises on company-level counter-insurgency and internal security in Afghanistan, and describes the principles of the Intelligence Cycle. It asserts that counter-intelligence is an operational necessity.

Medal Array from 1970s and 1980s

A medal array of the 1970s and 1980s, showing a BEM, South Atlantic Medal with Rosette, 1962 GSM (Northern Ireland), Accumulated Campaign Service Medal and Army Long Service and Good Conduct Medal. The Accumulated Campaign Service Medal was conceived to recognise multiple tours of duty in Northern Ireland with a qualifying period of 1,080 days.

Medal Array from 1990 and 2000s

GSM (N Ireland) with MID Oakleaf, 1993 First Gulf War. Former Yugoslavia (Bosnia and Yugoslavia, Accumulated Campaign Service Medal with three Bars (three years each) and LSGC.